AVERY WRIGHT

Floating Communities

The Emergence of Oceaniums

Contents

Thank You

Thank you for choosing "Floating Communities: The Emergence of Oceaniums". I hope this book has provided you with valuable insights into the potential of Oceaniums and their impact on society. I believe that by promoting environmental sustainability, social equity, and cultural exchange, Oceaniums can help create a better and brighter future for all. Thank you for joining us on this journey towards a more sustainable and equitable world.

Jonah Stone

1

Introduction

The world is facing a significant challenge in terms of sustainability, with increasing global concerns around climate change, environmental degradation, and the depletion of natural resources. With a growing population and the need for economic development, there is a growing demand for infrastructure and facilities that meet the needs of society while minimizing their impact on the environment. One area where this need is particularly evident is in the design and construction of sports stadiums, which typically require large amounts of energy and resources to build and operate.

The concept of Oceaniums offers an innovative and sustainable solution to this problem, by creating floating stadiums that utilize renewable energy and recycled materials, while also promoting marine biodiversity. The idea was first proposed by Vincent Callebaut Architectures, who designed the OCEANIUMS project as a biomimetic generation of floating and sustainable stadiums. In essence, Oceaniums are a cross between a boat and a stadium, creating community hubs that are intended for nomadism at sea.

The purpose of this book is to explore the emergence of Oceaniums as a new solution for sustainable stadium design, by examining the design principles, construction processes, and environmental benefits of this innovative concept. This book will also explore the potential impact of Oceaniums on society, economics, and geopolitics, and consider the challenges and opportunities of scaling up the concept to create a global network of floating communities.

The book is structured in twelve chapters, each of which examines a different aspect of the Oceaniums concept. The first chapter, which is this introduction, provides an overview of the book's purpose and structure, and explains the concept of Oceaniums in more detail.

The second chapter of the book focuses on the need for sustainable stadiums, and the environmental impact of traditional stadium design. This chapter will explore the benefits of sustainable stadium design and the potential impact of Oceaniums in this context.

Chapter three will examine the vision of Vincent Callebaut Architectures, and the OCEANIUMS project in more detail. This chapter will provide an overview of the features and benefits of Oceaniums, and how they differ from traditional stadiums.

Chapter four will explore the design principles of Oceaniums in more detail, explaining the biomimetic design approach used in the concept. The chapter will also describe the materials and technologies used in Oceaniums, including recycled materials and renewable energy systems.

Chapter five will examine the construction process of Oceaniums, including the challenges and opportunities of building floating stadiums. This chapter will also discuss the potential economic benefits of Oceaniums, including the potential for job creation and economic development.

Chapter six will explore the functioning of Oceaniums in more detail, describing how they generate and use energy, and the functions and activities that can take place in these floating stadiums.

Chapter seven will focus on the environmental benefits of Oceaniums, including their potential to promote marine biodiversity, and their use of recycled materials.

Chapter eight will examine the circular economy principles applied in Oceaniums, and how they can promote sustainable economic development. This chapter will also discuss the potential for Oceaniums to create new economic opportunities, such as the development of new technologies and industries.

Chapter nine will explore the potential for future development of Oceaniums, examining the challenges and opportunities of scaling up the concept to create a global network of floating communities.

Chapter ten will examine the social benefits of Oceaniums, and their potential to promote social change and community development.

Chapter eleven will focus on the geopolitical implications of Oceaniums, and how they could potentially impact global politics and international relations.

Finally, chapter twelve will provide a conclusion to the book, summarizing the key points made throughout, and reflecting on the potential for Oceaniums to create a sustainable future for society. Overall, this book aims to provide a comprehensive and insightful exploration of the

2

The Need for Sustainable Stadiums

The construction and operation of traditional stadiums have long been associated with significant environmental impacts, including the depletion of natural resources, the emission of greenhouse gases, and the production of waste. As the world's population continues to grow, and with increasing demand for sporting events and entertainment, it is essential to develop more sustainable approaches to stadium design and construction.

Traditional stadium designs typically rely on concrete and steel, which require significant amounts of energy and resources to produce. The construction process of these materials emits a large amount of greenhouse gases, contributing to climate change. The transportation of construction materials to the building site also adds to the carbon footprint of traditional stadiums.

Furthermore, traditional stadiums consume large amounts of energy during their operation, requiring electricity for lighting, ventilation, and air conditioning. This energy is often produced by burning fossil fuels, which also contributes to climate change.

In addition to their energy consumption, traditional stadiums also produce significant amounts of waste. This includes construction waste, food waste, and single-use plastics. Much of this waste is not recycled, leading to landfill and pollution.

Sustainable stadium design offers a solution to these problems, by reducing the environmental impact of stadium construction and operation. Sustainable stadiums are designed to minimize their energy consumption, reduce waste production, and promote the use of renewable energy sources.

One approach to sustainable stadium design is the use of recycled materials. Recycled materials can be used in the construction of stadiums, reducing the demand for new resources and minimizing waste. For example, the Levi's Stadium in California was built using over 80% recycled materials, including recycled steel and concrete.

Another approach to sustainable stadium design is the use of renewable energy sources. Solar panels and wind turbines can be used to generate electricity, reducing the reliance on fossil fuels. In addition, green roofs and walls can help to insulate buildings, reducing the need for heating and cooling systems.

Sustainable stadium design can also promote the use of public transportation to reduce the carbon footprint of fans traveling to and from the stadium. For example, the Mercedes-Benz Stadium in Atlanta was designed to promote the use of public transportation, with over 30 bus and train routes stopping within a mile of the stadium.

In addition to their environmental benefits, sustainable stadiums can also have economic and social benefits. Sustainable stadiums can create new jobs and industries, such as the development of renewable energy technologies. They can also promote community development by providing space for cultural events and activities.

Overall, the need for sustainable stadium design is clear. Traditional stadiums have a significant impact on the environment, and the development of more sustainable approaches is essential to reduce their carbon footprint and promote a more sustainable future. Sustainable stadium design offers a range of benefits, from reducing energy consumption and waste production to promoting economic and social development. The emergence of Oceaniums as a sustainable stadium design concept offers a unique and innovative approach to addressing these issues, creating floating communities that are sustainable, efficient, and resilient.

3

The Vision of Vincent Callebaut Architectures

The OCEANIUMS project, designed by Vincent Callebaut Architectures, represents a radical new approach to sustainable stadium design. Rather than constructing fixed, land-based stadiums that require significant amounts of energy and resources to build and operate, the OCEANIUMS project proposes creating floating stadiums that are powered by renewable energy and made from recycled materials. This innovative concept represents a significant step forward in sustainable design, providing a new model for how society can create infrastructure that meets the needs of the present without compromising the needs of future generations.

At its core, the OCEANIUMS project is based on the idea of creating floating communities that are self-sufficient, energy-efficient, and sustainable. These communities are designed to function as sports stadiums, cultural centers, and hubs of activity, all while promoting marine biodiversity and minimizing environmental impact. By utilizing renewable energy sources, such as solar and wind power, and using recycled materials, such as green algae and plastic waste, the Oceaniums concept is able to minimize its carbon footprint while maximizing its positive impact on the environment.

The key features of Oceaniums are their ability to navigate the ocean and to generate their own energy. Oceaniums use solar radiation and the strength of the prevailing winds to power their propulsion, while also utilizing the cold ocean currents of Labrador and the warm Gulf Stream to maintain temperature control. This allows Oceaniums to be self-sufficient, without relying on external sources of energy.

In terms of design, the Oceaniums concept is inspired by biomorphism, bionics, and biomimicry, which means that the organic geometries of the floating stadiums are designed to resemble natural forms found in the ocean. The structure of the Oceaniums is optimized for stability, resilience, and efficiency, drawing inspiration from the skeletons of cetaceans, coral reefs, and bioluminescent organisms.

The use of recycled materials is also a central feature of the Oceaniums concept. The Oceaniums project proposes using solid wood, recycled aluminum, green algae, and plastic waste from the 7th continent concentrated in the five oceanic gyres to build these floating stadiums. The plastic waste is transformed into new construction materials by feeding 3D printers that are connected to human-orchestrated artificial intelligence processors. This approach represents a significant step forward in circular economy principles, as waste and pollution are transformed into resources, and debts are transformed into solidarity.

The benefits of the Oceaniums concept are numerous. Firstly, by creating floating stadiums that are powered by renewable energy and made from recycled materials, the Oceaniums concept has the potential to significantly reduce its carbon footprint, helping to mitigate climate change. Secondly, by promoting marine biodiversity and using green technologies, Oceaniums can contribute to the preservation of marine ecosystems, helping to protect ocean life from the negative impact of human activity.

Furthermore, Oceaniums represent a new approach to stadium design, one that prioritizes sustainability, efficiency, and re-silience. The self-sufficient nature of Oceaniums means that they are able to operate in remote locations without relying on external sources of energy. This makes them ideal for use in areas where traditional stadiums would be impractical or impossible to construct.

Finally, Oceaniums offer significant economic and social ben-efits. By promoting the development of renewable energy technologies and the circular economy, Oceaniums can create new jobs and industries, contributing to economic development. In addition, by providing space for cultural events and activities, Oceaniums can promote community development and social cohesion.

The vision of Vincent Callebaut Architectures, as embodied in the OCEANIUMS project, represents a significant step forward in sustainable design. . By combining biomimetic design principles, circular economy practices, and renewable energy technologies, Oceaniums offer a regenerative and sustainable solution to the environmental and social challenges facing our planet.

As we look to the future, the potential of Oceaniums is immense. From providing sustainable and accessible sporting and cultural venues, to promoting marine biodiversity and sustainable tourism, to creating new economic opportunities and social mobility, Oceaniums have the potential to transform the way we think about infrastructure and its role in promoting a more sustainable and equitable world.

We invite you to join us on this journey towards a more sustainable future, and to embrace the vision of Oceaniums as a symbol of hope, innovation, and progress. By working together and supporting the development of sustainable infrastructure, we can create a better and brighter future for ourselves and for generations to come.

4

The Design of Oceaniums

The design of Oceaniums is based on biomimicry, a design approach that takes inspiration from nature to create sustainable and efficient solutions to human problems. The biomimetic design principles used in Oceaniums draw inspiration from natural forms and processes found in the ocean, resulting in a unique and innovative approach to stadium design.

One of the key biomimetic design principles used in Oceaniums is the optimization of structure. The structure of Oceaniums is inspired by the skeletons of cetaceans, coral reefs, and bioluminescent organisms, which have evolved over millions of years to be optimized for stability, resilience, and efficiency. By drawing inspiration from these natural structures, the design of Oceaniums is able to minimize the use of materials while maximizing structural integrity.

Another biomimetic design principle used in Oceaniums is the use of organic geometries. The organic geometries used in Oceaniums are inspired by natural forms found in the ocean, such as coral reefs, kelp forests, and bioluminescent organisms. These organic shapes are not only visually appealing but also contribute to the stability and efficiency of the floating stadiums.

The use of recycled and biosourced materials is also a central feature of the Oceaniums concept. Oceaniums are constructed using solid wood, recycled aluminum, green algae, and plastic waste from the 7th continent concentrated in the five oceanic gyres. The use of these materials helps to reduce the environmental impact of stadium construction, while also promoting the circular economy.

In addition to the use of recycled and biosourced materials, Oceaniums also utilize renewable energy sources. Solar panels and wind turbines are used to generate electricity, while the cold ocean currents of Labrador and the warm Gulf Stream are used to maintain temperature control. This approach not only reduces the carbon footprint of Oceaniums but also makes them self-sufficient and able to operate in remote locations.

The use of green technologies is another key feature of the Oceaniums design. For example, green roofs and walls are used to provide insulation, reducing the need for heating and cooling systems. In addition, natural ventilation systems are used to circulate air, further reducing energy consumption.

The technologies used in Oceaniums are also designed to be modular and scalable. This means that the stadiums can be easily assembled and disassembled, making them ideal for temporary events such as the World Cup and the Olympics. The modularity of the design also means that the stadiums can be adapted to different environments and locations, further increasing their versatility and sustainability.

Overall, the design of Oceaniums represents a significant step forward in sustainable stadium design. By drawing inspiration from natural forms and processes, Oceaniums are able to minimize their environmental impact while maximizing their efficiency and functionality. The use of recycled and biosourced materials, renewable energy sources, and green technologies also helps to promote the circular economy and reduce dependence on fossil fuels. The modular and scalable design of Oceaniums further increases their versatility and potential for use in a variety of contexts.

5

The Construction of Oceaniums

The construction of Oceaniums is a complex process that requires innovative design, engineering, and construction methods. The construction process for Oceaniums is unique in that it involves the assembly of floating stadiums that are designed to navigate the ocean and generate their own energy. In this chapter, we will explore the construction process for Oceaniums and discuss the challenges and opportunities associated with this process.

The first step in the construction process for Oceaniums is the design phase. The design of Oceaniums is based on biomimicry, a design approach that takes inspiration from nature to create sustainable and efficient solutions to human problems. The biomimetic design principles used in Oceaniums draw inspiration from natural forms and processes found in the ocean, resulting in a unique and innovative approach to stadium design.

Once the design has been finalized, the construction of Oceaniums can begin. The construction process involves the assembly of the various components of the floating stadium, including the hull, the superstructure, and the renewable energy systems. The hull of the Oceaniums is designed to be stable and efficient, drawing inspiration from the skeletons of cetaceans and coral reefs. The superstructure of the Oceaniums is designed to be visually appealing, incorporating organic geometries inspired by natural forms found in the ocean.

One of the major challenges of constructing Oceaniums is the use of recycled and biosourced materials. The Oceaniums project proposes using solid wood, recycled aluminum, green algae, and plastic waste from the 7th continent concentrated in the five oceanic gyres to build these floating stadiums. The plastic waste is transformed into new construction materials by feeding 3D printers that are connected to human-orchestrated artificial intelligence processors. This approach represents a significant step forward in circular economy principles, as waste and pollution are transformed into resources, and debts are transformed into solidarity.

Another challenge associated with constructing Oceaniums is the use of renewable energy sources. Solar panels and wind turbines are used to generate electricity, while the cold ocean currents of Labrador and the warm Gulf Stream are used to maintain temperature control. This approach not only reduces the carbon footprint of Oceaniums but also makes them self-sufficient and able to operate in remote locations. However, the installation and maintenance of these renewable energy systems can be complex and expensive.

Despite these challenges, the construction of Oceaniums also presents a range of opportunities. The use of recycled and biosourced materials and renewable energy sources not only reduces the environmental impact of stadium construction but also promotes the development of new industries and technologies. The modularity and scalability of Oceaniums also make them ideal for temporary events such as the World Cup and the Olympics, reducing the need for the construction of fixed, land-based stadiums.

In addition to the challenges and opportunities associated with the construction of Oceaniums, there are also ethical and social considerations that must be taken into account. The use of recycled and biosourced materials and renewable energy sources can promote the circular economy and reduce dependence on fossil fuels, but it is important to ensure that the production of these materials does not have a negative impact on local communities or ecosystems.

Overall, the construction of Oceaniums represents a significant step forward in sustainable stadium design. By utilizing biomimetic design principles, recycled and biosourced materials, and renewable energy sources, Oceaniums are able to minimize their environmental impact while maximizing their functionality and versatility. While there are challenges associated with the construction of Oceaniums, these challenges also present opportunities for innovation and the development of new industries and technologies.

6

The Functioning of Oceaniums

The functioning of Oceaniums is unique in that these floating stadiums are designed to be self-sufficient, generating their own energy and functioning as sustainable communities. In this chapter, we will provide an overview of how Oceaniums generate and use energy, as well as the functions and activities that can take place within these floating communities.

Oceaniums generate their own energy through the use of renewable energy sources such as solar and wind power. The solar panels and wind turbines used to generate electricity are integrated into the design of the stadium, ensuring that they are not only functional but also visually appealing. The cold ocean currents of Labrador and the warm Gulf Stream are also used to maintain temperature control, reducing the need for energy-intensive heating and cooling systems.

The renewable energy systems used in Oceaniums are designed to be modular and scalable, allowing for easy installation and maintenance. The modularity of the design also means that the renewable energy systems can be adapted to different environments and locations, further increasing their versatility and sustainability.

The functions and activities that can take place within Oceaniums are varied and diverse. As floating stadiums, Oceaniums are designed to host sports and cultural events, serving as hubs of activity and community gathering spaces. They can also function as research centers, promoting scientific research around marine biomimicry and other areas of environmental sustainability.

In addition to these functions, Oceaniums can also serve as centers of education and learning. They can be used to promote environmental education and awareness, teaching visitors about the importance of sustainability and marine biodiversity. They can also serve as training centers for sustainable practices, helping to promote the development of sustainable industries and technologies.

The self-sufficient nature of Oceaniums means that they are able to operate in remote locations without relying on external sources of energy. This makes them ideal for use in areas where traditional stadiums would be impractical or impossible to construct. The modularity and scalability of Oceaniums also make them ideal for temporary events such as the World Cup and the Olympics, reducing the need for the construction of fixed, land-based stadiums.

The use of recycled and biosourced materials in the construction of Oceaniums also promotes the circular economy, helping to reduce waste and pollution while creating new materials and resources. This approach not only reduces the environmental impact of stadium construction but also promotes economic development and the creation of new industries.

Overall, the functioning of Oceaniums represents a significant step forward in sustainable community design. By generating their own energy, promoting marine biodiversity, and utilizing recycled and biosourced materials, Oceaniums are able to minimize their environmental impact while maximizing their functionality and versatility. As floating stadiums and sustainable communities, Oceaniums represent a new model for how society can create infrastructure that meets the needs of the present without compromising the needs of future generations.

7

Oceaniums and Biodiversity

Oceaniums are designed to be havens of biodiversity dedicated to the flourishing of ocean flora and fauna. The design of Oceaniums draws inspiration from natural forms and processes found in the ocean, resulting in a unique and innovative approach to stadium design that prioritizes environmental sustainability and marine biodiversity.

One of the key environmental benefits of Oceaniums is their ability to promote marine biodiversity. By providing habitats for marine organisms and promoting the growth of ocean flora, Oceaniums are able to support the health and resilience of ocean ecosystems. The organic geometries used in Oceaniums are inspired by natural forms found in the ocean, such as coral reefs, kelp forests, and bioluminescent organisms. These organic shapes not only contribute to the stability and efficiency of the floating stadiums but also provide habitats for a variety of marine species.

The use of recycled and biosourced materials in the construction of Oceaniums also helps to reduce the environmental impact of stadium construction and promote the circular economy. By using plastic waste from the ocean as a construction material, Oceaniums are able to remove pollutants from the ocean and transform them into resources for sustainable development. This approach not only reduces the impact of stadium construction on the environment but also promotes the health and resilience of ocean ecosystems.

In addition to these benefits, Oceaniums also represent an opportunity to promote environmental education and awareness. By serving as centers of learning and research, Oceaniums can help to promote awareness of environmental issues and inspire new approaches to sustainable development. The use of renewable energy sources and green technologies in Oceaniums also serves as a model for how society can reduce its dependence on fossil fuels and promote the development of sustainable industries and technologies.

Overall, Oceaniums represent a new approach to stadium design that prioritizes environmental sustainability and marine biodiversity. By drawing inspiration from natural forms and processes found in the ocean, Oceaniums are able to minimize their impact on the environment while maximizing their functionality and versatility. As floating stadiums and sustainable communities, Oceaniums represent a new model for how society can create infrastructure that meets the needs of the present without compromising the needs of future generations.

8

Oceaniums and the Circular Economy

The concept of a circular economy, where waste is minimized and resources are reused, has become increasingly popular in recent years as a response to the environmental impact of our linear economic model. The circular economy principles applied in Oceaniums take this approach to the next level, by designing infrastructure that is not only sustainable but regenerative. In this chapter, we will explore the circular economy principles applied in Oceaniums and discuss the potential economic benefits of this innovative concept.

The circular economy principles applied in Oceaniums include the use of recycled and biosourced materials, the promotion of renewable energy sources, and the reduction of waste and pollution. By using materials such as recycled aluminum and plastic waste from the oceans, Oceaniums reduce the demand for virgin materials and promote the reuse of resources. This approach also helps to address the issue of plastic pollution in our oceans, turning a negative environmental impact into a positive resource for construction. Additionally, the use of renewable energy sources such as solar radiation and wind power reduces the dependence on fossil fuels and promotes the development of sustainable energy industries.

Another key circular economy principle applied in Oceaniums is the reduction of waste and pollution. The modular and scalable design of Oceaniums means that they can be adapted to different contexts and locations, creating opportunities for sustainable development and social change in diverse settings. This approach promotes the reuse of resources and the reduction of waste and pollution, creating a more sustainable and regenerative approach to infrastructure development.

The potential economic benefits of Oceaniums are significant. By promoting the circular economy, reducing waste and pollution, and supporting the growth of sustainable industries and technologies, Oceaniums can help to create new economic opportunities and spur innovation. The use of recycled and biosourced materials in the construction of Oceaniums also creates new opportunities for economic development and social mobility. Additionally, the promotion of renewable energy sources and sustainable industries creates new job opportunities and supports the growth of sustainable economies.

The circular economy principles applied in Oceaniums also have the potential to create new business models and revenue streams. For example, the use of recycled materials in the construction of Oceaniums can create new opportunities for waste management and recycling industries. The promotion of renewable energy sources can also create new opportunities for energy generation and distribution, as well as the development of new energy technologies.

However, the circular economy principles applied in Oceaniums also highlight the need for effective governance and regulation. The use of recycled and biosourced materials in the construction of Oceaniums raises questions about the regulation of these materials and the potential for environmental harm. Additionally, the promotion of renewable energy sources and sustainable industries requires effective policies and incentives to support their growth and development.

Overall, the circular economy principles applied in Oceaniums represent a new approach to infrastructure development that promotes environmental sustainability, social equity, and economic opportunity. By promoting the reuse of resources, reducing waste and pollution, and supporting the growth of sustainable industries and technologies, Oceaniums can help to create a more sustainable and regenerative future for society. The potential economic benefits of Oceaniums are significant, creating new opportunities for economic development, social mobility, and innovation. However, effective governance and regulation are essential to ensure that the development of Oceaniums is aligned with broader goals such as the UN Sustainable Development Goals.

9

The Future of Oceaniums

The concept of Oceaniums represents a new approach to stadium design that prioritizes sustainability, versatility, and marine biodiversity. As floating stadiums and sustainable communities, Oceaniums have the potential to transform the way society thinks about infrastructure, promoting a more circular and regenerative approach to development. In this chapter, we will examine the potential for future development of Oceaniums and discuss the challenges and opportunities of scaling up this innovative concept.

The potential for future development of Oceaniums is significant. The modular and scalable design of Oceaniums means that they can be adapted to different environments and locations, making them suitable for use in a wide range of contexts. The use of renewable energy sources, recycled and biosourced materials, and green technologies also promotes the development of sustainable industries and technologies, creating new opportunities for innovation and economic growth.

One of the key challenges associated with scaling up the Oceaniums concept is the need for investment in research and development. While the design principles used in Oceaniums are based on biomimicry and other established concepts, the implementation of these principles in the context of floating stadiums and sustainable communities requires innovative engineering and construction methods. Investment in research and development can help to overcome these challenges and promote the development of new technologies and materials that can make the construction of Oceaniums more efficient and cost-effective.

Another challenge associated with scaling up the Oceaniums concept is the need for collaboration and cooperation among different stakeholders. The construction of Oceaniums involves a wide range of actors, from architects and engineers to investors and policymakers. Effective collaboration among these actors is essential for ensuring that the construction of Oceaniums is sustainable, efficient, and effective. Collaboration can also help to ensure that the development of Oceaniums is aligned with broader goals such as the UN Sustainable Development Goals.

Despite these challenges, the potential for scaling up the Oceaniums concept is significant. As floating stadiums and sustainable communities, Oceaniums represent a new approach to infrastructure that prioritizes environmental sustainability and marine biodiversity. By promoting the circular economy, reducing waste and pollution, and supporting the growth of sustainable industries and technologies, Oceaniums can help to create a more sustainable and regenerative future for society.

In addition to these benefits, the scaling up of the Oceaniums concept can also have broader implications for society as a whole. By promoting sustainable development and environmental awareness, Oceaniums can help to shift societal attitudes toward sustainability and create a new model for development that meets the needs of the present without compromising the needs of future generations. The modular and scalable design of Oceaniums also means that they can be adapted to different contexts and locations, making them a powerful tool for promoting sustainable development and resilience in the face of environmental change.

Overall, the future of Oceaniums represents a significant opportunity for innovation and progress in the field of sustainable development. By promoting the circular economy, supporting marine biodiversity, and promoting sustainable industries and technologies, Oceaniums can help to create a more sustainable and regenerative future for society. While there are challenges associated with scaling up the Oceaniums concept, these challenges also present opportunities for collaboration, innovation, and progress.

10

The Impact of Oceaniums on Society

The impact of Oceaniums on society extends beyond their environmental and economic benefits. As floating stadiums and sustainable communities, Oceaniums also have the potential to promote social change, inspiring new approaches to community development, cultural exchange, and social equity. In this chapter, we will provide an overview of the social benefits of Oceaniums and discuss their potential to promote social change.

One of the key social benefits of Oceaniums is their ability to promote cultural exchange and community building. As floating stadiums, Oceaniums are designed to host a wide range of cultural and sporting events, serving as hubs of activity and community gathering spaces. This creates opportunities for people from different cultures and backgrounds to come together, share their experiences, and build new connections. The modular and scalable design of Oceaniums also means that they can be adapted to different contexts and locations, creating opportunities for cultural exchange and community building in diverse settings.

Another social benefit of Oceaniums is their potential to promote social equity and inclusion. By providing sustainable, affordable, and accessible community infrastructure, Oceaniums can help to address social and economic disparities and promote greater social equity. The use of recycled and biosourced materials in the construction of Oceaniums also promotes the circular economy, creating new opportunities for economic development and social mobility.

In addition to these benefits, Oceaniums can also promote social change by inspiring new approaches to sustainable development and community design. The innovative design principles used in Oceaniums draw inspiration from natural forms and processes, challenging conventional approaches to infrastructure and promoting new models of sustainable development. The use of renewable energy sources, green technologies, and recycled and biosourced materials also promotes the development of sustainable industries and technologies, creating new opportunities for innovation and economic growth.

The potential for Oceaniums to promote social change is significant. As floating stadiums and sustainable communities, Oceaniums represent a new approach to community development that prioritizes sustainability, inclusivity, and cultural exchange. By providing accessible and sustainable community infrastructure, promoting social equity, and inspiring new approaches to sustainable development, Oceaniums can help to create a more just, equitable, and sustainable society.

Overall, the impact of Oceaniums on society extends beyond their environmental and economic benefits. By promoting cultural exchange, community building, social equity, and social change, Oceaniums represent a new model for community development that meets the needs of the present without compromising the needs of future generations. The modular and scalable design of Oceaniums also means that they can be adapted to different contexts and locations, creating opportunities for social change and innovation in diverse settings. As floating stadiums and sustainable communities, Oceaniums represent a powerful tool for promoting a more just, equitable, and sustainable future for society.

11

The Global Implications of Oceaniums

The potential for Oceaniums to have a global impact is significant. As floating stadiums and sustainable communities, Oceaniums represent a new approach to infrastructure that prioritizes environmental sustainability, social equity, and cultural exchange. In this chapter, we will discuss the potential for Oceaniums to have a global impact and examine the geopolitical implications of this innovative concept.

One of the key global implications of Oceaniums is their potential to promote environmental sustainability and reduce the impact of human activity on the planet. By using renewable energy sources, recycled and biosourced materials, and green technologies, Oceaniums can help to reduce greenhouse gas emissions and promote a more sustainable and regenerative approach to development. The circular economy approach used in Oceaniums also promotes the reuse of resources and the reduction of waste and pollution, helping to create a more sustainable and equitable future for all.

Another global implication of Oceaniums is their potential to promote cultural exchange and understanding. By providing hubs of activity and community gathering spaces, Oceaniums can help to break down cultural barriers and promote greater cross-cultural understanding. This can help to reduce tensions and promote peace and cooperation between different countries and regions, creating opportunities for collaboration and shared progress.

The geopolitical implications of Oceaniums are also significant. By promoting sustainable development and reducing the impact of human activity on the planet, Oceaniums can help to address some of the key geopolitical challenges of our time, including climate change, environmental degradation, and social inequality. The modular and scalable design of Oceaniums also means that they can be adapted to different contexts and locations, creating opportunities for sustainable development and social change in diverse settings.

However, the geopolitical implications of Oceaniums also raise important questions about governance, regulation, and accountability. The construction of Oceaniums involves a wide range of actors, including architects, engineers, investors, and policymakers, and effective collaboration among these actors is essential for ensuring that the development of Oceaniums is aligned with broader goals such as the UN Sustainable Development Goals. The use of recycled and biosourced materials in the construction of Oceaniums also raises questions about the regulation of these materials and the potential for environmental harm.

Overall, the global implications of Oceaniums are significant, representing a new approach to infrastructure that prioritizes environmental sustainability, social equity, and cultural exchange. By promoting sustainable development, reducing the impact of human activity on the planet, and breaking down cultural barriers, Oceaniums represent a powerful tool for addressing some of the key geopolitical challenges of our time. However, the development of Oceaniums also raises important questions about governance, regulation, and accountability, highlighting the need for effective collaboration and alignment with broader goals and priorities.

12

Final Thought

In this book, we have explored the concept of Oceaniums, float-ing stadiums and sustainable communities that prioritize envi-ronmental sustainability, social equity, and cultural exchange. We have discussed the need for sustainable stadiums and the benefits of sustainable stadium design, as well as the vision of Vincent Callebaut Architectures and the design, construction, functioning, impact on society, and global implications of Oceaniums.

One of the key points that emerged from our discussion is that Oceaniums represent a new approach to infrastructure that pro-motes the circular economy, reduces waste and pollution, and supports the growth of sustainable industries and technologies. The modular and scalable design of Oceaniums also means that they can be adapted to different contexts and locations, creating opportunities for sustainable development and social change in diverse settings. Additionally, Oceaniums promote marine biodiversity, cultural exchange, and social equity, creating new opportunities for innovation and progress.

The construction of Oceaniums involves a wide range of actors, including architects, engineers, investors, and policymakers, and effective collaboration among these actors is essential for ensuring that the development of Oceaniums is aligned with broader goals such as the UN Sustainable Development Goals. The use of recycled and biosourced materials in the construction of Oceaniums also raises questions about the regulation of these materials and the potential for environmental harm. These challenges highlight the need for investment in research and development, effective governance, regulation, and account-ability, and collaboration among different stakeholders.

In conclusion, Oceaniums represent a significant opportunity for innovation and progress in the field of sustainable devel-opment. By promoting the circular economy, reducing waste and pollution, and supporting the growth of sustainable indus-tries and technologies, Oceaniums can help to create a more sustainable and regenerative future for society. The potential for Oceaniums to promote cultural exchange, social equity, and environmental sustainability is significant, representing a new approach to infrastructure that meets the needs of the present without compromising the needs of future generations. The modular and scalable design of Oceaniums also means that they can be adapted to different contexts and locations, creating opportunities for sustainable development and social change in diverse settings.

13

Bonus Chapter: The Role of Education in Promoting Oceaniums

Education is a crucial component in promoting the concept of Oceaniums and encouraging the adoption of sustainable practices. In this chapter, we will explore the role of education in promoting Oceaniums and discuss the ways in which education can help to create a more sustainable future for society.

Firstly, education can help to raise awareness about the environmental challenges facing our planet, including the impacts of climate change, pollution, and the loss of biodiversity. By teaching students about these issues, educators can encourage them to think critically about the world around them and inspire them to take action to protect the environment. Through educational programs, students can learn about the benefits of sustainable practices, such as recycling, reducing waste, and using renewable energy sources.

Secondly, education can play a key role in promoting the circular economy principles applied in Oceaniums. By teaching students about the importance of reducing waste and promoting the reuse of resources, educators can help to instill a culture of sustainability and responsible resource management. Educational programs can also promote the development of new business models and revenue streams that prioritize the circular economy, creating new opportunities for economic development and social mobility.

Thirdly, education can promote cultural exchange and understanding, helping to create a more diverse and inclusive society. By teaching students about different cultures and promoting intercultural dialogue, educators can help to foster a sense of global citizenship and encourage students to think beyond their own communities. This can help to promote empathy, tolerance, and understanding, creating a more peaceful and equitable world.

Finally, education can help to promote the UN Sustainable Development Goals, providing a framework for sustainable development and encouraging action to create a more sustainable and equitable world. By teaching students about the UN Sustainable Development Goals, educators can inspire them to think about how they can contribute to creating a better future for all. Through educational programs and initiatives, students can learn about the potential of Oceaniums to promote environmental sustainability, social equity, and cultural exchange, and can become advocates for this innovative concept.

Education plays a vital role in promoting the concept of Oceaniums and encouraging the adoption of sustainable practices. Through educational programs and initiatives, students can learn about the environmental challenges facing our planet and the potential of sustainable practices to address these challenges. By promoting the circular economy principles applied in Oceaniums, educators can help to create new opportunities for economic development and social mobility. Educational programs can also promote cultural exchange and understanding, creating a more diverse and inclusive society. Finally, education can promote the UN Sustainable Development Goals, providing a framework for sustainable development and encouraging action to create a more sustainable and equitable world.

14

Glossary

Biomimetic – Design that is inspired by nature and mimics its principles.

Biosourced – Materials that come from renewable sources and are biodegradable.

Circular Economy – An economic model that prioritizes the reuse of resources, reduces waste and pollution, and promotes regenerative growth.

Cultural Exchange – The sharing of ideas, beliefs, and practices between different cultures.

Environmental Sustainability – The ability to meet the needs of the present without compromising the ability of future generations to meet their own needs.

Marine Biodiversity – The variety of plant and animal life in the world's oceans and seas.

Modular Design – Design that is made up of separate modules or components that can be easily combined or replaced.

Renewable Energy – Energy that comes from sources that are naturally replenished, such as solar, wind, and tidal power.

Scalable Design – Design that can be easily expanded or reduced in size to meet changing needs.

Social Equity – The fair distribution of resources and opportunities in society.

Sustainable Development – Development that meets the needs of the present without compromising the ability of future generations to meet their own needs.

UN Sustainable Development Goals – A set of 17 goals adopted by the United Nations to promote sustainable development, social equity, and environmental sustainability.

Waste Management – The process of collecting, transporting, processing, and disposing of waste materials in a safe and environmentally responsible manner.

Waste Reduction – The process of minimizing waste and promoting the reuse of resources.

Wind Power – Energy generated by the movement of air across the earth's surface, usually through the use of wind turbines.

About the Author

Avery Wright is an enigmatic figure who is an author in the fields of AI, Technology, and the Arts. A combat veteran of the US Army, Avery has almost two decades of experience in the IT industry, which has given them a unique perspective on the intersection of technology and society.

As an author, Avery has published a range of books on topics such as the future of AI, the role of drones in modern warfare, and the medicinal properties of mushrooms. Their writing often explores the cutting-edge of technology and how it is changing the world around us. Avery's work is notable for its depth and insight, as well as its ability to make complex topics accessible to a broad audience.

Away from the world of writing, Avery is a private individual who values their privacy. Despite this, they remain a voice in the tech industry and beyond. Whether sharing their thoughts on the latest developments in AI or commenting on the state of the world, Avery's perspective is always worth listening to.

You can connect with me on:

🌐 https://sirexodia.wixsite.com/avery-wright

𝕏 https://twitter.com/AveryWrightAI

𝐟 https://www.facebook.com/profile.php?id=100089987171726

🔗 https://www.amazon.com/author/averywrightai

Subscribe to my newsletter:

✉ https://sirexodia.wixsite.com/avery-wright

Also by Avery Wright

Avery Wright's work spans a range of topics, from the cutting-edge of AI and technology to the ancient practice of Taoism and the art of chess. Their books are notable for their depth, insight, and ability to make complex topics accessible to a broad audience. With almost two decades of experience in the IT field and a background as a combat veteran, Avery brings a unique perspective to their writing that is both informative and thought-provoking. Whether you are interested in exploring the frontiers of technology or deepening your understanding of the human experience, Avery's books are a must-read.

Mastering Midjourney AI - The Beginner's Handbook
Mastering Midjourney AI: The Beginner's Handbook is a comprehensive guide for beginners looking to learn about the Mid-journey AI platform and how to use it for image generation. The book covers a range of topics, including understanding Midjourney AI's parameters and settings, using URLs for image inspiration, adjusting image quality, and more.
https://www.amazon.com/dp/B0BV8PGDXT

Mastering Prompt Engineering for Chat-GPT: A Beginner's Guide (ChatGPT Masterclass)

Mastering Prompt Engineering for ChatGPT: A Beginner's Guide" is a comprehensive guide to prompt engineering for ChatGPT. Whether you're a beginner or an experienced user, this book will provide you with the principles and best practices to effectively guide ChatGPT towards generating accurate and relevant responses in a variety of use cases. From understanding the capabilities and limitations of ChatGPT to crafting effective prompts and evaluating their effectiveness, this book covers all the essential topics and strategies you need to master prompt engineering for ChatGPT.

https://www.amazon.com/dp/B0BWYTJ8B8

Unleashing ChatGPT's Power: Navigating the Digital Realm

"Unleashing ChatGPT's Power: Navigating the Digital Realm" explores the complex and sophisticated technology behind one of the most innovative and exciting technological advancements of our time. Through in-depth discussions of the underlying technology that powers ChatGPT, its ability to process natural language, and its expanding knowledge base, readers will gain a deeper understanding of the ways in which artificial intelligence and natural language processing are shaping the future of society.

https://www.amazon.com/dp/B0BVSYQ77H

Chat GPT: A Digital Journey Begins

ChatGPT: A Digital Journey Begins" is the first book in an exciting series that takes readers on a journey into the world of artificial intelligence and natural language processing. Through the development and evolution of ChatGPT, readers will gain an in-depth understanding of the complex and sophisticated technology that powers this groundbreaking model. From the creation of the neural network to the potential applications of ChatGPT in various fields, this book provides a comprehensive and fascinating overview of one of the most innovative and exciting technological advancements of our time.

https://www.amazon.com/dp/B0BVQB5M91

The Tao of Inner Peace

The Tao of Inner Peace is an introduction to the ancient Chinese philosophy and religion of Taoism. This book explores the core teachings of Taoism and how they can be applied in everyday life to find inner peace and harmony. The book covers a range of topics, including the concept of Tao, the Yin-Yang philosophy, the Tao Te Ching, living in harmony with nature, the Tao of relationships, and the Tao in action. With practical guidance and advice, this book will help readers cultivate a more peaceful and fulfilling life by adopting a Taoist approach to everyday living."

https://www.amazon.com/dp/B0BVMJ83DZ

www.ingramcontent.com/pod-product-compliance
Lightning Source LLC
Chambersburg PA
CBHW070750220526
45467CB00018B/1874